Good Health & Long Life

Meanderings of a Mead Maker

Brian P. Dennis

NORTHERN BEE BOOKS

Good Health & Long Life

Meanderings of a Mead Maker

ISBN 978-1-904846-76-5

Published by Northern Bee Books, 2011
Scout Bottom Farm
Mytholmroyd
Hebden Bridge HX7 5JS (UK)

Design and Artwork, D&P Design and Print
Printed by Lightning Source UK

Good Health & Long Life

Meanderings of a Mead Maker

Dedication:

I thank June, my wife, for supporting my beekeeping activities over many years and Rachel & Hannah, my daughters, for tolerating their eccentric father.

I dedicate this book to my grand-daughter Hesper in the hope that bees & beekeeping continue to flourish and she finds an interest in life that is as rewarding as beekeeping has been to me.

Good Health & Long Life
Meanderings of a Mead Maker

A Czar once wanted to discover the secret of long life, so he held an inquiry to find out which section of his population lived longest. He found it was the beekeepers and immediately presumed that honey was the life-giving food. A more detailed inquiry to find out how much honey it was necessary to eat revealed the alarming fact that all the beekeepers were far too poor to eat any honey at all, and they lived on the refuse that was left over from the harvest. Perhaps it was the mead that kept them going.

Kenneth K. Clark in *Beekeeping.*

Mead is probably the oldest alcoholic drink known. The word for mead occurs in nearly all Indo-European languages. There is good reason to believe that mead was known some 12,000 years ago. It was certainly very popular in Anglo-Saxon times and in 'polite society' right up to the end of the 17th century.

The extraordinary long life of the ancient Britons has frequently provoked comment and speculation on the part of historians. Among the early Romans, Plutarch once observed "The Britons only begin to grow old at the age of 120" and when Pliny visited the British Isles he reported "These islanders consume great quantities of honey brew". Pollio Romulus wrote to Julius Caesar, when over 100 years old, that he enjoyed a full sex life, which he attributed to drinking copiously of Welsh Mead.

Virgil and Homer wrote about mead in glowing terms. The Greeks had a definite mead-making session. The mead was matured and kept for an orgy called a Dionysia. What they did on these occasions, under the influence of mead, must be left to the imagination. Hippocleides, for example, having drunk too much mead on his wedding night, stood on his head on the dining table, stark naked, waving his legs in the air while he sang a merry song. His father refused to let him take his bride!

Scandinavians expected to quaff mead in heaven out of the skulls of their enemies. On earth, the Vikings were wont to consume at least half a dozen horns of mead during a meal.

Ethelstan, a Saxon King in England during the 10th century, expressed his delight at the copious supply of mead provided upon a visit to his warlike Aunt, Ethelfleda, *Lady of the Mercians.* It would appear that the mead during this period was extremely potent – it is recorded that in monasteries, the monks were limited to a sextarium (about a pint) of mead to be shared between six at dinner and half that quantity at supper.

In more recent times, Samuel Pepys wrote in his diary for 1666 "Dined with two or three of the King's servants ... I ... had methyglin ... which did please me mightily". Methyglin, or spiced mead, was much liked by Queen Elizabeth I and she gave very detailed instructions as to which herbs should be used as flavourings - her recipe has survived to this day.

In the same century, Sir Kenelm Digby wrote: "The Meathe (mead) is singularly good for a consumption, stone, gravel, weak-sight and many more things. A chief burgomaster of Antwerp used for many years to drink no other drink than this and though he were an old man, he was of extraordinary vigour, had always a great appetite, good digestion and had every year a child.".

The *birds and the bees* would seem to be involved in the history of mead! The Moors wedding celebrations were sex-orgies where the guests were made drunk on mead because they believed honey to be a love stimulant. Indeed, in this country the word *honeymoon* comes from the practice of drinking mead during the month long celebrations which followed better-class weddings. In some parts it was the custom to send the bride to bed and then fill the bridegroom with mead until he could no longer stand. He was then carried to bed alongside his wife and it was believed he would then sire a son that night. If successful the maker of the mead was complimented on its quality. Is it possible that mead is the youth elixir of antiquity?

In most inns mead was sold along with ale and cider. The Anglo-Saxon word for mead was *alu* (cf. *ale*). Much mead was beer strength and made with hops. There was trouble in the British Army during the Napoleonic Wars when the strength of the troops' mead was reduced from 6% to 4%. The slow decline of mead as the national drink can be traced back to the Norman Conquest for in the wake of the invasion came the first wine traders.

Imports of grape wine rocketed when Henry II married Queen Eleanor of Aquitane, whose possessions included Bordeaux. The dissolution of the monasteries by Henry VIII put an end to monastic brewing. In the 18th century sugar replaced honey. The importation of cheap spirits and the Industrial Revolution, which involved the loss of rural skills such as home wine making, completed the decline.

The claims made for mead are almost as many and as widespread as those for honey. One would imagine that a daily dose of honey would cure a wooden leg! While there may be some truth in the therapeutic value, they would seem to be very much exaggerated. The chemical analysis of honey indicates nothing that substantiates such claims, although 181 substances have been found in it to date. Scientific proof is sadly lacking - *evidence* is subjective and really comes down to *If you believe it does you good, then it probably will!* The same cause and effect argument is seen when claiming that beekeepers rarely suffer rheumatism because they are stung by bees. The syllogism is:

- Beekeepers get stung.
- Beekeepers don't suffer from rheumatism.
- Therefore, bee sting prevent/cure rheumatism.

It is more likely that the regular exercise involved in moving heavy supers etc. in the sun and fresh air keeps the body free of *the rheumatics*. Still if you are paying someone a lot of money to apply bees to your skin in order to cure pain, you would want to believe it was doing you good. The same sort of argument occurs in the first quote - the *refuse* undoubtedly refers to the cappings which the peasant Russian beekeepers ate or turned into mead. It may or may not have been this which led to their longevity. However, I digress from the main topic - mead.

The latest fossil evidence suggests that bees existed and were producing honey *50 million years ago*. Homo sapiens' mere 5 *million years* of evolution explains, perhaps, why bees have such a well organized society and why we are still fighting each other. I digress again ...

Early man *hunted* for honey as he did for many other foods (and as some people do today). A painting made in a rock shelter in the mountains of eastern Spain in Mesolithic times, probably about 7,000 B.C., survives to show how this was done. The combs were broken off from the nest and eaten - a balanced diet of wax, honey, pollen, brood and probably a few dead bees as well. Presumably little attempt was made to store honey. Gradually there occurred a shift from hunting for honey to keeping bees in purpose made hives made from local materials - in this country, bees were kept in straw baskets or *skeps*. The earliest known record of keeping bees in hives and harvesting their honey dates from 2,400 B.C. in Egypt. The practice of selecting the skeps with most honey - the heaviest ones - at the end of the year and killing the bees by placing the skep over a pit containing burning sulphur was developed. The honey would have been strained and stored in various containers, perhaps not completely sealed, thus allowing the stored honey to absorb moisture. Some of the honey is likely to have been unripe i.e. containing an excessive amount of water, and would have been fermented by wild yeasts which abound in the air. So by his attempts to preserve honey to use over a period, man probably introduced himself to a fermented alcoholic drink and found it to his liking! In this way, mead production is believed to have begun - the making of alcoholic beverages for man's comfort and pleasure, before the grape took over.

How then can we make some mead? Most books on wine making and beekeeping contain recipes for mead. Like most hobbies (including beekeeping), the process may be as simple or as complicated as you like - with all the associated apparatus you can afford. In recent years, there has been a resurgence of interest in home brewing and wine making. There has developed an industry to provide equipment and materials. No longer do we have to follow out-of-date recipes and rely on 'natural fermentation' or float rafts of baker's yeast on toast in open buckets or brew *tonic ale*. And, so long as you don't sell your results, it is perfectly legal. For this article we need not get too complicated, but we can make use of modern knowledge and the ease with which we can now obtain equipment.

I am surprised by the few beekeepers of my acquaintance who make mead and those who say they don't like the taste, considering the range of tastes that can be produced. Some of them, I suspect, have made mead in the past and produced something unpleasant - or even vinegar (= sour wine)! Or they may have purchased commercial mead which, in my experience, is always sickly sweet and strongly flavoured. Many years ago I bought a bottle of mead from a large wine shop in Soho, London. What was I doing there? Mind your own business - I did say **many** years ago! I asked the assistant if it was sweet. "Of course it is", he replied, "it is made from honey." I did not point out that dry wines are made from sweet grape juice. Why are commercial meads usually so sweet?

The type of honey used determines the flavour and bouquet of the finished product. Light coloured honey is best for making dry light meads with subtle flavours. Use dark honeys for strong flavoured sweet meads. Most beekeepers will use their own honey but if you use bought honey avoid eucalyptus honey from Australia - it makes a mead with a most unpleasant flavour. Flavours in honey masked by the sweetness become more noticeable when the sugar is fermented. The quantity of the honey determines the alcoholic strength and final sweetness.

When making mead, yeast is added to the honey dissolved in water. During fermentation the yeast feeds on the sugar in the honey and splits it into carbon dioxide and alcohol. The carbon dioxide gas bubbles away leaving the alcohol behind (fortunately!). Yeast also needs nutrients and acid to keep it growing and working. These are lacking in honey and must be added. Tannin is also needed to give the mead astringency and to assist clarification. During fermentation the liquor is susceptible to spoilage by micro-organisms, ever present in the air, the most important of which are the vinegar bacteria which convert alcohol into acetic acid (vinegar). To avoid competing with *Sarsons Vinegar*, utensils must be sterilised and air must be excluded during fermentation using an air-lock.

Let's make some mead!

Ingredients:

3 - 3½ lb honey.
½ oz citric acid.
½ tsp tannin (or ½ cup black strong tea).
2 tsp yeast nutrient.
Wine yeast (Maury yeast has been specially selected for mead but a General Purpose Yeast will be suitable).
2 tsp yeast nutrient & ¼ tsp yeast extract (e.g. *Marmite*) to provide vitamin B.
Water to 1 gal.

(Specific Gravity [SG] approx. 1.100 = potential alcohol 13.4%)

Method:

You can obtain your equipment and ingredients from any wine making supplier.

Warm the honey in three times its own volume of water, stir to dissolve (avoid burning the honey), bring just to the boil and simmer for a couple of minutes. Remove the scum. Do not boil fast as many desirable substances will be evaporated, causing loss of flavour and bouquet.

When cool, transfer to a 1 gallon glass jar (demijohn) previously well rinsed with hot water. Bring the remaining water to the boil and when cool add to the dissolved honey. Add the yeast, nutrient, tannin and acid. Fit an air lock (or plug the neck of the jar with cotton wool) and leave in a warm place. When fermentation is complete (when there are no more bubbles and it has begun to clear), siphon using a length of plastic tubing (or carefully decant) the mead into a clean jar leaving the sediment behind. When another deposit has formed, siphon again. When it no longer throws a sediment and is clear, bottle. If necessary, filter or add wine finings.

The above recipe should produce a *dry mead* containing about 13% alcohol. If the finished mead tastes rather sweet, delay bottling until you are sure fermentation has finished to avoid burst bottles. A *medium mead* would need about 4 lb honey and a *sweet (or sack) mead* 4½ lb.

Sultanas give extra flavour, body and smoothness to mead and nourish the yeast. Rinse 12 oz sultanas in warm water and chop or mince. Ferment on the pulp, stir daily, and strain after 10 days.

Your mead will probably be drinkable after a year. Having made mead, don't be impatient to drink it - there is no comparison between young mead and the matured article. Brother Adam of Buckfast Abbey recommended maturing mead in sound oak casks for a full seven years before bottling. I have never achieved such perfection. At least hide a couple of bottles to mature and make some more. **5 gallons lasts almost twice as long as 1 gallon!** And of course, if you are a beekeeper, you will enter a bottle of mead in the *National Honey Show* and your local Association Show. Having entered a bottle of mead at a local show, I approached the judge and told him that the mead he had awarded First Prize was *awful.* He looked rather surprised until I explained that the mead was mine! His reply was "You should have tasted the others"!

If you are a beekeeper and wish to use the honey remaining in cappings, you need to measure the amount of honey dissolved in your liquor. The old method was to float a new laid egg in the dissolved honey and when only a piece of shell the size of an old sixpence was showing, the amount of honey was correct. Nowadays, one can purchase an instrument called a hydrometer which is easy to use and much more reliable. A hydrometer measures the *specific gravity* (SG) of the sugar solution.

Place the cappings in a suitable container and add cold water. Stir to dissolve the honey, allow to stand a while and then strain. Take a hydrometer reading and adjust with honey or water to give the required starting gravity. More honey will increase the specific gravity, more water will lower it. Proceed as in the recipe above.

2 lb honey in 1 gal gives SG 1.060, potential alcohol 7.8%.
3 lb honey in 1 gal gives SG 1.090, potential alcohol 12%.
4 lb honey in 1 gal gives SG 1.120, potential alcohol 16.3%.

4 lb honey added to 1 gal = 3 lb. **in** 1 gal.

Dry Mead:
Starting SG 1.085-1.105. Finish SG 0.990-1.000.

Medium Mead:
Starting SG 1.105-1.120. Finish SG 1.000-1.005.

Sweet Mead:
Starting SG 1.120-1.130. Finish SG 1.005-1.015.

If you add one equal quantity of water the gravity (not the Specific Gravity) will be reduced by half e.g. from 180 to 90 (or from SG 1.180 to 1.090).

When making a sweet mead it is a good plan to add half the total honey at the outset, and the remainder in 4 oz lots each time the SG approaches 1.000.

Mead is fermented honey and water. By adding other ingredients you may produce interesting variations. Originally, of course, herbs were added for medicinal purposes - so they say! Spices were added, I suspect, to mask the taint of vinegar etc. A famous drink of the well-to-do was known as *pyment* or *piment*. This was a mixture of grape juice (sometimes already fermented) and honey.

Very often spices were added and the brew was then called *hippocras*. This had a great many variations with names associated with the Church such as *Pope*, *Cardinal* and *Bishop* and was, sometimes served hot in cold weather. Clerics were often criticised for their excessive taste for them. In 817 a local synod at Aix la Chapelle tried to ban the clergy from drinking spiced wines. *Mulsom* was wine made into a long sweet drink with honey and water, its name being given to *mulled wine.* Stone bottles were filled with mulled mead to warm the occupants of the bed - and then the contents were drunk!

Braggot, Braggon or *Bracket* was a mixture of ale or beer, honey and often spices, very popular in the 13th century. Braggot was a drink peculiar to Cheshire, Lancashire and Yorkshire and was much favoured in Wales. It was often called *Varii* and was said to be better to use on ships than wine. Chaucer, when writing of the carpenter's wife in *The Miller's Tale,* says:

Hir mouth was as swete as braggot or the meeth,
Or hord of apples leyd in hay or heath.

Mothering Sunday was known as Bracket Sunday in Lancashire, when this drink was served to the men and women *in service* visiting home.

However, here are the traditional variations:

Pyment: grape juice and honey. *Hippocras:* pyment and herbs.
Cyser: apple juice and honey.
Morat: mulberry juice and honey.
Melomel: fruit juice (other than apple, grape or mulberry) and honey.
Metheglin: dry mead with herbs and spices.
Sack mead: sweet mead.

Hippocras is named after Hippocrates, the Greek physician and 'father of medicine'. The cloth bag that held the herbs was called the *hippocratic sleeve.*

Wales and Cornwall are famed for their honey and mead. Metheglin (or metheglin, theglin, medd) means medicine in both languages. The Welsh *meddyglyn* and the Cornish *medheklyn* derive from the Latin *medicus* and the Old English *hlynn* meaning liquor. In Wales, *Meddiginiaeth* became synonymous with a remedy or cure by means of *medd* and the medical man was called *meddig* i.e. one who cures with the *medd.* Similarly, *meddwi* meant 'drinking excessively', *meddw* meant 'being drunk' and *meddw wyf fi* meant 'I am drunk'!

Sack probably derives from *sack* sherry, from the Spanish word *sacar* meaning *to export,* rather than *seco (sec, siccus)* meaning *dry.*

Thus Water boils, parboils and mundifies,
Clears, cleanses, clarifies and purifies.
But as it purgeth us from filth and stink,
We must remember that it makes us drink,

Metheglin, Bragget, Beer and headstrong Ale,
(That can put colour in a visage pale) ...

John Taylor (*'Water Poet'*) *1580 -1653.*

Clara Furness in *Honey Wines and Beers* also gives the following:

Clarre:	pyment
Alicante wine:	morat
Myritis:	bilberries and honey
Rhodomel:	rose petals and honey
Miodomel:	hops and honey

These variations ferment more readily and mature more quickly. However the addition of spices may cause hazes which will have to be removed by filtration or fining with proprietary finings. If you already make wine, replace the sugar in the recipe with honey using 1 lb for ¾ lb sugar to allow for the water content of the honey.

Metheglin I - Queen Elizabeth's Recipe

Take of sweet briar leaves and thyme each one bushel, rosemary half a bushel, bay leaves one peck. Seethe these ingredients in a furnace full of water (probably not less than 120 gallons) boil for half an hour, pour the whole into a vat and then when cooled to a proper temperature (approx. 75°F.) strain. Add to every 6 gals. of the strained liquor a gallon of fine honey and work the mixture together for half an hour. Repeat the stirring occasionally for two days, then boil the liquor afresh, skim it till it becomes clear and return it to the vat to cool, when reduced to a proper temperature (approx. 80°F.) pour it into a vessel from which fresh ale or beer has just been emptied, work it for three days and tun. When fit to be stopped down, tie up a bag of beaten cloves and mace (about half an ounce of each) and suspend it in the liquor from the bung hole. When it has stood for half a year it will be fit for use.

Whelm your kiv!

When Charles Butler describes how metheglin was made for Good Queen Bess he states: *When the mead is clear you must have in readiness a kiv of new ale which as soon as you have emptied suddenly whelm it upside down and set it up again."*

Modern Version

Make a gallon of mead as described above. Suspend in the finished mead a muslin bag containing ½ oz rosemary, ½ oz bay leaves, ½ oz thyme and ¼ oz sweet briar. Taste the mead daily until the flavour is to your liking and remove the herbs.

Metheglin II

4½ lb dark honey.
1 oz each of mace, cloves, cinnamon, bruised ginger.
Thin rinds of 1 lemon & 1 orange.

Simmer together, strain, cool.
Add yeast, nutrient, acid, tannin. Ferment.
Needs long maturation.

Pyment

1 pint white grape concentrate.
2 lb heather honey.

¼ oz citric acid.
Yeast, nutrient, tannin.

Combine ingredients and ferment.

To make *Hippocras*:

1. Add ¼ oz cinnamon at start of fermentation.
2. Add 1 knob of bruised root ginger and the juice and peel (no pith) of 1 small orange, boiled in a pint of water for 20 minutes and strained over the honey etc.
3. Add **one** of the herbs or flowers from the following:
   Parsley, Marjoram, Cowslip (4 oz fresh, 1 tsp dried).
   Mint, Sage, Caraway seeds, Meadowsweet, Lemon thyme, Elderflowers, Balm (2 oz fresh, 1/2 tsp dried).
   Mace (1 oz fresh, 1/4 tsp dried).

A mixture of any of the above herbs can be used - do not exceed 2 - 4 oz . Ferment on the pulp for 4 days, stir daily, strain, etc.

Sachets of herbs are available for making mulls and herbal teas. Experiment with these and whole and powdered spices from the kitchen (nutmeg, peppercorns, coriander, citrus peel, etc.). Infuse the herbs in the mead and remove when the strength of flavour is sufficient. Powdered herbs should be placed in a muslin bag.

Sprigs of thyme, rosemary or fennel standing in a bottle of brilliantly clear mead makes an unusual gift.

Cheat's Melomel or Cyser

Supermarkets stock a wide range of inexpensive fruit juices. A range of melomels can easily be made using 1 litre of juice (plus yeast, nutrient, tannin and 1 tsp citric acid) and 3 lb honey. Cyser can be produced using 1 - 3 litres of apple juice with 3 lb honey and a little less acid and tannin (¼ tsp acid and a pinch of grape tannin). Added body can be achieved by boiling two bananas in sufficient water to cover & adding the strained infusion.

A hybrid between cyser and pyment can be made using 1 lb honey, 1 pint of white grape concentrate and 4 pints of apple juice. 2 lb of raisins could be substituted for the grape concentrate.

Many concentrates are available from wine making shops. Follow the instructions for making a wine but substitute honey for the sugar.

Recipes show variation in acid and tannin content, depending to some extent on the ingredients. I have not, you will notice, tried to convert recipes to metric quantities. I have tried to be consistent, but recipes are always a guide - experiment. You can measure the acid content, but taste is usually sufficient. Our ancestors may have had their fair share of failures - but the long history of mead suggests they had their successes - without scientific knowledge or equipment.

If you like the end result - drink it and make some more! I once made a wine that I thought was undrinkable. I donated it to a friend's party! The party was to entertain some French visitors - they thought the wine excellent and a great aperitif!

Earlier in this book I referred to several variations such as metheglin & melomel etc. The Greeks had their own variations. *Thalassiomel*, for example, was made from sea water and honey - a sort of sailors' beer drunk shortly after fermentation had ceased. *Kykeon* sounds even worse being composed of oil, wine, cheese and meal. I have not included recipes for these concoctions! I did mention rhodomel, which was according to one reference served at nude bathing parties. If chilled slightly it would make a pleasant drink in the garden on a hot summer's day - dress would be optional!

Rhodomel

1 pint heavily scented rose petals (lightly pressed down).
2 lb honey.
¼ oz tartaric acid + ¼ oz malic acid (or ½ oz citric acid). Yeast & nutrient.

Dissolve honey in 6 pints of warm water and pour over the rose petals - add nutrient and acid. Add 2 Campden tablets. When cool add the yeast and ferment on the rose petals for 3 days. Strain and transfer to a demijohn - ferment to dryness. Rack without splashing and add a Campden tablet. Top up the jar. Rack again at 4 months later and then six months, each time without splashing and with the addition of a Campden tablet. The rhodomel should then be drinkable and may be sweetened slightly with ¼ lb sugar.

I also mentioned *morat* and *sack mead*.

William the Conqueror reduced the number of court banquets from four a day down to one! This must have seriously reduced the consumption of morat and pigment. Pigment, as the name suggests, was said to "colour the stomach", which is not surprising since it contained large amounts of ginger, cinnamon, nutmeg and cloves. Morat, however, was a light, medium sweet, table wine with a silky undertone. In order to achieve the silkiness a large dose of sulphite (Campden tablets) is added at the start of the fermentation to delay fermentation for anything up to a fortnight. When the yeast does get going it makes glycerine until the sulphite is used up, after which normal fermentation proceeds. The sulphite protects the must against bacterial infection.

Morat

3 lb mulberries.
2 lb honey.
Yeast & nurient.

Dissolve honey in 4 pints of water. Add the juice of the pressed mulberries. Leach the mulberries with 2 pints of boiling water and strain onto the mixture. Make volume up to 1 gal and add 5 Campden tablets. After 48 hours add the yeast and nutrient. When fermentation is completed, rack without splashing, add 1 Campden tablet and top up. Rack again at 4 monthly intervals. Drinkable at 1-2 years.

Sack or sweet mead was once a favourite drink mentioned by Shakespeare and earlier writers. The recipe for Sack 1 originally had the dissolved honey boiled for 2 hours, which would result in loss of flavour. Don't add more fennel or the flavour will be unpleasantly strong.

Shakespearean Sack 1

3 or 4 fennel roots.
3 or 4 sprays of rue*.
41b honey.
2 tsp citric acid.
Yeast & nutrient.
2 gal water.

Wash the roots and leaves and boil in water for 45 mins. Strain and add the honey. Bring to the boil and simmer for a few minutes, remove scum. Make volume up to 2 gal and transfer to demijohns. When cool, add the yeast and nutrient. Rack as necessary. Should be fit for drinking after a year.

Shakespearean Sack II

1 small piece of fennel root.
1 small sprig rue*.
4 lb medium honey.
1 tsp citric acid.
Sherry yeast & nutrient.
1 gal water.

Bring the honey to the boil in the water with the fennel & rue. Boil for 2 or 3 minutes, remove scum. Strain and add acid, yeast & nutrient when cool.

The two recipes show the variations to be found in recipes especially quantities and boiling times. It is a matter of personal taste - experiment. Finished meads can be blended to achieve a desired taste.

Sack mead should be matured for at least three years and will improve for up to ten years. They are strong meads and should be consumed with caution.

N.B. Wear gloves when picking rue. Rue can cause hypersensitivity to sunlight & a mild burn or blister may result.

The reason rue has declined in popular use is that while using small amounts of the herb is beneficial, rue is poisonous in large amounts and can cause violent stomach upset, skin irritation as well as photosensitivity.

The following sack mead contains hops so, according to Clara Furness, this is a *Miodomel.*

Miodomel

4lb honey.
¼ oz hops.
8 oz sultanas.
¼ oz citric acid.
Sherry yeast & nutrient. 1 gal water.

Bring **3 lb honey** and the chopped sultanas to the boil in 6 pints of water and skim. Boil the hops in 2 pints of water for 10 minutes. Strain and add to honey & sultana mix. Add yeast, acid & nutrient when cool and ferment for 4-5 days. Strain off sultanas. When gravity drops to 1.005 add 4 oz honey dissolved in some of the mead - repeat until the remaining 1 lb honey has been added. If the mead tastes dry, sweeten to taste. The initial flavour of the hops will disappear when the mead has matured.

In my later article on honey beers I refer to heather ale about which Robert Louis Stevenson wrote:

From the bonny bells of heather,
They brewed a drink lang-syne,
Was sweeter far than honey,
Was stronger far than wine.
They brewed it and they drunk it,
And lay in blessed swound,

For days and days together,
In their dwellings underground.

Heather Ale

1 gal flowering heather tips (lightly pressed down).
½ oz hops.
¼ oz root ginger.
2 tsp citric acid.
1 lb heather honey.
Ale yeast & nutrient.

Boil the heather tips, hops, bruised ginger and honey for 15 minutes in 7 pints of water. Strain into bucket, add acid, nutrient and, when cool, yeast. Cover well and skim for 3 days. Ferment to dryness. Bottle in beer bottles and prime with a level teaspoon of sugar per quart. Should be drinkable after a fortnight - then you can *"lay in blessed swound together"*!

Many old recipes go to great lengths to obtain purified water, making one realize how much we take for granted in modern society. The use of egg whites and egg shells, sand or charcoal filters or prolonged boilings and settlings were required to remove extraneous matter. No wonder clean spring water was often recommended.

A recipe c. 77 AD for *hydromel* (Greek *hydromeli* = water-honey) recommends that *rain-water should be stored for five years. Some who are more expert,* it continues, *use rain-water as soon as it has fallen, boiling it down to a third of the quantity and adding one part of old honey to three parts of water, and then keeping the mixture in the sun for 40 days after the rising of the Dog-star. Others pour it off after nine days and then cork it up. With age it attains the flavour of wine.* (*Natural History* by Pliny the Elder.)

According to Clara Furness, hydromel is the name given in France to a wine made of grape juice enriched with honey. W Herrod-Hempsall states that hydromel is a French name for a liquor consisting of honey diluted with water, fermented or unfermented. Attila the Hun is said to have died through drinking too freely of hydromel at his wedding feast!

Hydromel, at least in the above recipe, is simply honey and water with no other additions. However, like many old country wine recipes, fermentation probably relied on airborne wild yeasts to promote natural fermentation. Other recipes for hydromel include yeast. Sir Kenelme Digby (1699) gives the following recipe:

Hydromel As I Made It Weak For The Queen Mother:

18 quarts of spring water.
1 race of ginger.
4 cloves.
1 quart of honey.
1 spoonful of ale yeast.
1 sprig of rosemary.

The Queen Mother was Henrietta Maria who was exiled to France after the execution of her husband King Charles I.

The Greeks were not the only ones who had a name for honey brews. A drink from Korea called *Misshu* is a spirit made from distilled mead, which is drunk with milk and poppy juice. It is said to make the drinkers very optimistic and cheerful and to give fantastically wonderful dreams! Across the Indo-Aryan world, a drink called *Soma* made from the juice of a plant mixed with milk, barley meal and honey was believed to give visions and ensure immortality. *Madhu, Haoma* and *Amrita* were similar drinks.

Conditum is another curious drink made from fermented honey and pepper and said to be a digestive remedy.

Conditum Paradoxum - c. 100 B.C. to 300 A.D.

6 sextarii honey.
4 oz crushed pepper.
3 scruples mastic.
20+ sextarii wine.
1 dram spikenard or bay laurel leaves.
1 dram saffron.
5 drams roasted date stones (previously softened in wine). Crushed charcoal for filtering.

Bring honey and 2 sextorii of wine to the boil, add a dash of cold wine, remove from heat and skim. Repeat 2 or 3 times, rest until next day and skim again. Add pepper, mastic, spikenard or laurel, saffron and date stones. Add remainder of wine. Add charcoal to clarify.

Sextarius = 1.5 pints = 0.9 litre.
Scruple = 1.3 g.
Dram = 1/8th oz (Apothecary measure).
Mastic = resin from mastic tree (*Pistacia lentiscus*) used nowadays in chewing gum.

Russia is one of the countries where mead and other honey drinks remained popular.

Lipez is one of these. Although it is produced like a beer it is of wine strength and should not be drunk from tankards.

Lipez

- 2 lb lime blossom honey.
- 2 lb barley.
- ½ lb malt extract.
- ½ oz hops.
- ¼ oz citric acid.
- Beer yeast and nutrient.

Boil honey and hops in 6 pints of water for 15 minutes and strain over the barley, malt extract and acid. Cover and leave for 12 hours. Strain and add the yeast and nutrient. Skim any scum that appears. When SG drops to below 1.008, bottle in beer bottles and mature tor 6 weeks.

References to mead, sex and long-life are many and world-wide and have given some examples. The Scots have a saying that mead drinkers have as much strength as meat eaters. In Scotland centuries ago, a wasting disease was treated by a concoction known as *Athol Brose*, which contained heather honey and Scotch whisky, taken little and often.

Athol Brose

- 3 heaped dessertspoons oats (not *Instant*).
- 8 oz cream.
- 8 oz whiskey.
- 3 dessertspoons sherry.
- 2 dessertspoons clear honey.

Cover oats with water and soak overnight. To the liquid (*brae*) add the cream, whisky, sherry and honey, stirring after each addition. Bottle and refrigerate for five days. Remove an hour or so before use, let it thaw a little, pour into a bowl, stir and ladle into glasses. This does not keep indefinitely.

The impression one has of winters during Victorian times is of the country's inhabitants battling through snow and biting winds, perhaps travelling on top of an open stage coach, to be revived on arrival at their destination with bowls of steaming hot mulled wine. The best description of such a scene is to be found in Charles Dicken's *Pickwick Papers*. Nowadays winters are less severe and we have central heating! But the following *winter warmers* are still worth making to greet your guests at Christmas and at other times.

All mulled drinks should be hot but not boiling. Traditionally wine (and ale) was mulled by inserting a heated poker kept especially for the purpose. The secret is to heat & stir the ingredients in a saucepan, tasting and making additions as necessary, until the contents begin to boil round the edge. It should then be served, putting a metal spoon in each glass to prevent the glass cracking. Gentlemen developed their own recipes and you can follow their example and experiment. Dr Johnson, for example, was a great maker of *Bishops*. In his day, if you were known in a tavern, you could make your own Bishop. Think of all those tasting sessions you will have to undertake! If you make your own wine (apart from mead), you can substitute elderberry wine for port, etc.

Oxford Nightcaps

Three cups of this a prudent man may take;
The first of these for constitution's sake,
The second to the lass he loves the best,
The third and last to lull him to his rest.

Bishop 1

Heat together a bottle of claret, a sliced orange, 2 tablespoons of honey, 4 cloves and ½ pint of water. When just boiling, add a wine glassful of Curacao and one of brandy. Pour into glasses and grate a little nutmeg on top.

Dr Johnson dried the orange peel afterwards and grated it into a glass of port, insisting it was good for indigestion. In London, they made Bishops with oranges rather than lemons.

*... in summer it may be iced **Bishop** is often made of Madeira in England, and is perfumed with nutmegs, bruised cloves, and mace. It ought, however, to be made of old generous Bourdeaux wine or it fails of its purpose as a tonic liqueur ... When this compound is made of Bourdeaux wine, it is called simply **Bishop**; but according to a German amateur, it receives the name of **Cardinal** when old Rhine wine is used; and even rises to the dignity of **Pope** when imperial Tokay is employed.*

'The Cook and Housewife's Manual: a Practical System of Modern Domestic Cookery and Family Management' by Mistress Margaret Dods. 1829.

*If a dry red wine, like claret is used it is called a **Cardinal** while if Champagne is used it is known as a **Pope ... Lawn Sleeves** ... is made exactly as a **Bishop** except that madeira or sherry is substituted for port and three glasses of hot calves-foot jelly are added.*

'Making Mead' by Bryan Acton & Peter Duncan.

Bishop II

Insert cloves into the rind of an orange and roast in an oven. Put small but equal amounts of cinnamon, mace, allspice and a piece of ginger into a saucepan. Add ½ pint of water and boil until reduced by half. Bring a bottle of port to the boil and apply a lighted taper to burn off a portion of the alcohol. Add the roasted orange and spices and heat. Rub some lump sugar on the rind of an unroasted orange and add together with the juice. Grate in some nutmeg and sweeten with honey.

Negus

In the early 18th century, Colonel Francis Negus was MP for Ipswich. One cold night in the House of Commons, he had the idea of heating the port they were drinking and diluting the strength to make the debates more endurable. The result is a simple Bishop.

Heat together a bottle of red wine, ¼ pint of water, ½ lb of honey, 6 cloves and the pared rind and juice of a lemon. Add a tot of brandy or rum and some grated nutmeg.

A *caudle* or *caudel* was a warm, spiced drink of wine or ale, thickened with egg yolks, similar to egg nog. This was popular in Elizabethan times and was used as a cold cure, as a winter nightcap and for insomnia. In the days of unheated bedrooms, the drink was taken to bed in a quart mug and held between the legs to warm the body. It was then drunk to warm the body inwardly ensuring a good night's sleep! It was also given to women in labour and invalids or friends enjoying a *rere-soper* or illicit late night meal. From *caudle* come the words *coddle* and *mollycoddle*. When thickening with eggs, it is important not to boil the mixture or it will curdle and have lumps - best served immediately.

Flemish Caudle - c. 1375

Water.
White wine.
Egg yolks.
Salt.
Verjuice (a sour juice made from unripe fruits such as grapes or crab apples. Substitute mixture of cider vinegar and grape juice).

Put a bit of water to boil. Take egg yolks beaten without the whites, temper (= mix) with white wine, thread into your water, and stir very well so that it does not curdle. Add some salt and move it to the back of the fire. Some add just a bit of verjuice.

The following version does not contain eggs:

2 pints brown ale.
¼ lb honey.

1 tbsp oatmeal. Pinch nutmeg. Juice of a lemon.
1 wineglass whisky or rum.

Pour ale over honey & oatmeal and stand in front of the fire or place in a slow oven for a couple of hours - cover to avoid loss of alcohol. Stir and strain. Add nutmeg, lemon juice & whisky or rum.

This is similar to mulled ale - the following recipe for which *does* contain eggs!

Mulled Ale -1860

1+ pint ale.
3 eggs.
Grated nutmeg. Sugar.
1 tbsp brandy. Toast sippets.

Boil a pint of ale with nutmeg and sugar (honey could be substituted). Beat up the eggs and mix with a little cold ale: then gradually add the hot ale and pour backwards and forwards from one vessel to the other several times, to prevent it curdling. Warm, and stir till it thickens, then add a tablespoonful of brandy, and serve hot with toast.

Obarne was scalded mead similar to mulled ale, much used in Russia.

A *posset* (or *poshet* or *possot*) was a warm drink of spiced, sweetened, milk, curdled with wine or ale. Most recipes contain a large proportion of cream and eggs. The following is one of the simplest. Sweeten with honey instead of sugar.

A Plain Ordinary Posset

1 pint of milk.
4 spoonsful ale. (Spoonful = 1 dram 6 grains!)
2 spoonsful Sack wine (= Sherry).
Sweeten to taste.
Bring the milk to the boil - allow to cool. Pour onto the wine, ale and sugar. Let it stand.
From *The Closet ... Opened*, 1699.

Syllabub is also made with sweetened curdled milk or cream. A curd forms on the top while the clear liquid settles to the bottom. Syllabub is traditionally passed around and drunk out of a special spouted jug or cup. Some recipes require the cow being milked directly into the syllabub! This produced a frothy head - pouring the milk from a height should produce the same effect.

Syllabub

1 quart apple cider.
Milk.
Sweeten to taste.
½+ pint sweet cream.
Grated nutmeg.

Sweeten the cider, add grated nut [illegible] n you have added what quantity of milk you think proper, pour half a pint or more, in proportion to the quantity of syllabub you make, of the sweet cream you can get all over it." The quantities in many old recipes appear somewhat imprecise - the way my mother used to cook!

From *The First American Cookbook, etc.* by A Simmons. 1796.

Melpop

1 lb honey.
3 pieces crushed ginger.
1 lemon, sliced.
1 gal water.
Yeast.

Pour boiling water on ingredients. Ferment.

Capillaire

Maidenhair fern (*Adiantum capillus-veneris*) was exported from Ireland in the 18th & 19th century as a garnish and as the main ingredient in a flavouring syrup called *capillaire*, which was added to various fruit drinks or barley water as a tonic.

Many different recipes evolved and it later became a Parisian spiced drink, which included orange flower water. Rather than giving a recipe which requires boiling ferns and adding orange flower water, the following appears slightly easier!

Into a wide-necked container put ½ lb honey, ¼ lb. preserved ginger, 2 oz candied lemon peel, juice of 2 lemons and the peel of one and 2 wine glasses of blackcurrant juice. Add a gallon white wine and cover. After a month, strain (filter if necessary). Serve chilled or warmed.

In the early 17th century, the custom of *wassailing* apple trees was performed by farmers and their workers to ensure a good crop next year. *Wassail* or *Wass Hal* means "Be thou of good health". The time of this ceremony varied from area to area - in some places it took place on Christmas Eve and in others on Twelfth Night. One custom was to make a strong drink, float raisins and set fire to the surface (brandy was cheap!). Children would play *Snapdragon* in which they picked the flaming raisins off the surface. When they had all been consumed, everyone gave a *wassail* toast to each other using the remaining drink. In West Sussex, it was the custom to wassail orchards and apiaries. The song used on the latter occasion was as follows:

Bees, of bees of Paradise,
Does the work of Jesus Christ,
Does the work that no man can,
God made man and man made money,
God made bees and bees make honey,
God made great men to plough and sow
And God made little boys to tend the rook and crow.

Hurra!

Wassail Bowl

Boil together 1 pint water, 1 cup honey, 4 cloves and 3 sticks cinnamon for 5 minutes. Add two thinly sliced lemons and stand for 7-8 minutes. Add a bottle of medium dry red wine and heat slowly to just below boiling point. Serve very hot.

It would seem that our ancestors were more adventurous with their drinks than we are today used to getting our needs off the supermarket shelf. I have not tried all the recipes but writing this has whetted my appetite! I'll finish with a recipe for *Lamb's Wool*, a spiced ale traditionally drunk on New Year's Day in Scotland:

Heat four gallons of beer with three pounds of honey, four teaspoonsful of grated nutmeg, 2 oz ginger and the juice of four lemons. Strain and serve. You may need to invite some friends to help drink this quantity! CHEERS!!

HEAT.
USE A MILK
BOTTLE AS
A TRIAL JAR.
FERMENT.
BOTTLE.
SYPHON.

"NOT A LOT OF PEOPLE KNOW THAT"

"There was a gentleman here yesterday", he said, "a stout gentleman, by the name of Topsawyer ... he came in here ... ordered a glass of ale - would order it - I told him not - drank it and fell dead. It was too old for him. It oughtn't to be drawn, that's the fact."

Charles Dickens - *'The Waiter'*

I am like Autolycus in Shakespeare's The Winter's Tale, *"a snapper up of unconsidered trifles"* or as my more critical friends call me *"a fund of useless knowledge"*. In particular, I collect snippets of beekeeping lore as avidly as any philatelist collects stamps! Did you know, for example, that clergymen live longer than most? It is too soon to know whether 'clergywomen' will equally benefit from their calling. But beekeepers live longer than clergymen. It follows, therefore, that beekeeping clergymen must have an enviable life-span. As Michael Cain says "Not a lot of people know that" - and like most generalisations, such a conclusion may contain at least a grain of truth.

Nowadays, in this health conscious, health obsessed age, much of what brings pleasure and relaxation in our increasingly stressful lifestyles is bad for us. The fact that experts disagree and their collective wisdom keeps changing doesn't help. I have known women as thin as lathes on slimming diets, convinced they were overweight and overweight men who, after years of being sedentary and eating, drinking and smoking to excess, took up jogging - and dropped dead from heart attacks. So much *for mens sana in corpore sano.* A more hedonistic lifestyle is my aim - my ambition is to be shot at the age of 90 by an irate husband! After all, an expert is only 'x' the unknown quantity and 'spurt' a drip under pressure ....

This set me thinking about earlier times when life was hard. I have a snippet that refers to the author's great-grandfather who kept bees, made his own mead, and smoked an ounce of thick twist a day since the age of 13. This great ancestor was against things - mainly Temperance Societies! It was only after he had outlived all his opponents that he decided, at the age of 107, that he had had enough and quietly departed. Incidentally, mentioning Temperance Societies brings to mind the origin of the word teetotal. A proponent of total abstinence from intoxicants had a stutter. When he addressed meetings, he referred to t-t-t-total abstinence. Hence, teetotal! However, did the drinking of mead contribute to great-grandfather's longevity?

As I wrote in the previous chapters, the claims made for mead are almost as many and widespread as those for honey. But honey defies scientific analysis - although many components have been identified, including essential minerals and vitamins, there is nothing in such abounding quantity to explain the great benefit that honey and mead seem to promote. There is almost a perverse satisfaction in having a mystery trapped in a bottle of excellent mead. The only scientific evidence that is often cited is that of the Colorado bacteriologist who transferred typhoid bacteria into honey to see if it would act as a suitable medium for breeding further quantities of the bacteria. To his surprise, within 48 hours the bacteria were dead. In a later experiment, some honey was impregnated with the combined bacteria of typhoid fever, paratyphoid, enteritis, dysentery, bronchopneumonia, septicaemia and peritonitis. Within five days not a single bacterium remained alive.

Until about 1750, honey was the principal sweetening agent. It was not until the Industrial Revolution, when machinery was invented to process sugar cane from the West Indies, that sugar ceased to be a luxury enjoyed only by the rich. The creation of plantations and slave labour and the opening up of the West Indies made sugar cheaper and available to most. In 1747, in France, Marggraf discovered that sugar beet had a high sugar content - this was exploited for sugar production during the Napoleonic Wars in an effort to beat the British blockade (although, in this country, it was not until the early part of this century that beets were used for sugar production). Honey was abandoned and mead making, already undermined by hopped beer and the importation of cheap spirits and wines, went into decline.

In more recent times, it has been suggested that many ailments can be attributed to, or worsened by, a large intake of white sugar. Much evidence is cited in books promoting the health-giving and curative properties of honey, but most appears to be folklore rather than scientific research. But Professor Jung has pointed out that if one finds a belief in several unconnected cultures at different times in history, one should be wary of dismissing it out of hand. Some claims can be easily verified. Massaging the scalp with honey is said to cause the shiniest pate to sprout hair. Anyone like to check this one out? However, beekeepers - even the less hirsute - appear to be a healthy lot. A curious fact arising from all this is that quite small quantities of honey or mead are needed to produce large health benefits. A couple of glasses of mead a day will assist one's health - once one has been drinking thus for about a year. **I drink mead for my health - it is medicinal.**

I hope you had a go at making some mead following my previous article. While that is maturing, why not try some quicker honey brews? If you already make beer you will know how to go about it. If you don't, talk to someone who does or read a book on home brewing. The process is simple enough but home-made beers are conditioned (made gassy) by adding a small amount of sugar to each bottle. If beer is bottled too soon or too much priming sugar is added, the result can be like Vesuvius or a fire extinguisher on opening the bottle. It can also result in burst bottles which, of course, is **highly dangerous** if you are holding it at the time. When the fermentation is complete and no more bubbles are rising and the beer is beginning to clear, transfer to an other container to clear further. A hydrometer will enable you to have greater control. Honey beer contains less sugar than mead and consequently takes a shorter time to ferment - usually fermentation takes about a week.

Do use beer bottles designed to withstand pressure - you may have to purchase bottled beer and drink the contents to obtain the bottles!! Only food grade plastics should be used. Everything should be sterilised. Keep back some of the hops and add 5 minutes at the end of boiling the *wort* (= the dissolved honey liquor). Boil hops in as much of the water as your container will hold - vigorously. Bottles need to be sealed with crown caps. These and other equipment and ingredients can be purchased from winemaking and brewing suppliers.

Honey Ale

Ingredients:
4 oz honey.
1 lb sugar.
1 oz ground ginger.
2 fl oz lime juice.
½ oz citric acid (or juice of 3 lemons).
1 gal water.
General Purpose wine yeast.
Yeast nutrient.

Method:

Boil 4 pints of water with the ginger for half-an-hour. Pour onto the honey, sugar, lime & lemon juices in a suitable container. Stir to dissolve. Add 4 pints cold water and when cool, add the yeast & nutrient. Ferment, closely covered, until fermentation has finished. Siphon from sediment (*racking*) and then bottle and prime.

The Celtic name for Britain was *The Honey Isle of Beli.* The Celts were as keen on honey brews as were the Saxons and appear to have distilled mead. Dr Howells of Oxford writing in the 18th century states that the Celtic druids and bards used to drink mead before engaging in speculations - unfortunately, he does not support his assertions with any references. The Celts sometimes added ingredients to their mead to impart magical properties. For example, they fermented a mixture of honey and the juice of either the hazel or birch tree which they believed endowed them with superhuman strength - as a result of which they probably went about pillaging their neighbours! But one of the good things about mead is that it is not a depressive, unlike many other drinks, and they undoubtedly pillaged with great joy and exuberance!

Irish saints were partial to a drop of mead - no *Guinness* and no sex. St Findian, who lived on bread and water all week, used to eat salmon and drink mead on Sundays. Saint Brigitte, emulating Christ's miracle at Cana, turned a great vat of water into mead when the King of Leinster visited and they ran out of drink. Unfortunately she did not leave a recipe for this 'Instant Mead'.

Heather ale was known to the Picts. About 1500, the Principal of Aberdeen University wrote: *The Pichtis maid of this herbe, namit hadder, ane richt delicious and hailsum drinke.* When the Pictish kingdom fell to the Scots, heather ale seems to have fallen from fashion, although it was known to the Danes who invaded Ireland in the 9th century. Three Danes, a father and his two sons, are believed to be to have been the last to know the recipe. They were captured by the Irish after the battle of Clontarf and refused to exchange the recipe for their lives. What was the secret?

Celtic Ale

Ingredients
7 pints birch sap.
3 lb honey.
½ oz malic acid.
¼ oz tartaric acid.
(or ¾ oz citric acid - 2 level tsp.)
Tannin (or cup of strong black tea).
White wine or General Purpose yeast.
Yeast nutrient.

Method

Obtain the birch sap. In early May, drill tree with diameter over 20 cm (8") - 1 cm hole, 2 cm deep (just beyond bark), about 30 cm above ground. Insert a length of plastic tubing - plug neck of collecting jar with cotton wool. It may take about a week to obtain sufficient sap - plug the hole when enough sap has been obtained or the tree may bleed to death. Add the honey and other ingredients (except the yeast), stir well to dissolve the honey and add 2 Campden tablets. 24 hours later add the yeast and ferment to dryness. Rack and top up with water. Mature 1-2 years. Drink in small glasses if you don't want trolls dancing on your head in the morning - this is powerful stuff! This is more like a wine and does not require priming.

Mead was drunk on great occasions such as a wedding feast but for everyday use a low-powered mead-ale was made. In the late 18th century, there came population growth and a drift away from the countryside to the towns. To feed the growing population, much greater acreage of land was used for cereal production and barley for beer became cheaper than honey for mead or ale. In *Wassail in Mazers of Mead* G.R. Gayre says "The knowledge of malting spread from the ancient warm, temperate world to the North, and the use of malted barley (as a cheap substitute) took the place of honey". When ale from honey and ale from malt were both available, that made from honey was considered superior. It was only the increasing cheapness of malt ale which caused honey ale to go out of general production. The 4th century Greek explorer Pytheas wrote in his diary "Wherever grain and honey flourish men use it for making drink".

Incidentally, originally ale was malt liquor without hops and beer was hopped. Soldiers returning from the Hundred Years War (1338-1453) could well have acquired the taste for hops and created a demand for the drink to which they had been accustomed in France - *bere* or *biere*, ale flavoured with hops. In this country, ale was flavoured with herbs such as nettles, rosemary, alecost, ground ivy, etc. Beer is referred to by name in documents dating from A.D. 900-970 but subsequently disappears from the language. During Anglo-Saxon times the distinction was made between ale and beer - a physician advised his patient to cease drinking beer, although he was allowed to drink *ale* and *mead!* By the tenth century the word *beor* was used as an equivalent to *idromellum*, an inferior type of mead.

Anglo-Saxon Beer

Ingredients
1½ lb honey.
½ oz hops.
¼ oz citric acid
Beer yeast & nutrient.
1 gal water.

Method

Bring the honey to the boil and skim. Add the hops and simmer for 15 minutes. Strain. Add the lemon juice and nutrient. Make up to 1 gallon and, when cool, add the yeast & nutrient. When fermentation has finished, rack, bottle and prime (drinkable after about 8 weeks).

Barm, according to my dictionary, is "the yeasty froth on fermenting malt liquors (Old English bearm)" - hence *barmy* is slang for insane (i.e. full of barm, frothing, excited). Another name for yeast was godesgood. An ancient law decrees that any brewery must provide barm to anyone wishing to brew beer at home – as far as I am aware, **the law has never been repealed**. Take your container to your local brewery or buy a beer yeast from your supplier. You can keep some yeast from each brew to ferment the next.

The Romans brought hops to Britain for use as a vegetable rather than for brewing. After the Romans left there is no mention of hops for several centuries. Their return is recorded in the 17th century rhyme thought to be referring to A.D. 1520:

Hops. Reformation, bays and beer
Came into England all in one year.

Other versions list *Hops and turkeys, carp and beer and Turkeys, carps, hops, pickerel and beer.* One author states that the rhymes are inaccurate since hops were grown and used well before the Reformation and pickerel was known in Medieval times. Pickeral is dialect for small or young pike. There is mention of brewers of beer (birra) as distinct from ale in Hythe in 1419. Hops were later grown in Kent by Flemish settlers. However, in 1424 brewers using hops were accused of adulteration. In 1484 brewers of ale petitioned that *hoppes, herbs and the like should not be used in ale, only licour, malt and yeste'.* Consequently, brewers of beer were fined. Eventually, in 1493, *beer* brewers were recognized as a Guild and towards the middle of the 16th century both ale and beer were brewed with hops. But even as late as 1512 the authorities of Shrewsbury were prohibiting the use of the *'wild, pernicious weed, hops'* and the following year the brewers of Coventry were ordered not to use hops. In 1530, Henry VIII ordered his brewer in Eltham not to add hops to the ale.

Anglo Saxon Beer II

Ingredients

1 lb honey.
1 oz hops.
¼ oz citric acid.
Beer yeast & nutrient.
Water to 1 gal.

Method

Boil honey and hops for 45 minutes. Strain. Add citric acid and nutrient etc.

Similar recipes suggest:

1 lb honey, ½ hops, juice of 1 lemon, 30 min. boil.
1½ lb honey, 1 oz hops, no acid, simmer 1 hour.
1½ lb honey, 1 oz hops, no acid, 30 min. boil.

This is described as *Honey Botchard*, which is probably the same as Botchet. Botchet was used in the place of beer, mainly in Yorkshire and Lancashire.

Let's lift our mug o'botchet, to our mistresses charms ...
An quaff a stoup o'Betty's brew
Botchet frothing.

The Tippler's Song (1700) sung at *Mell* (harvest suppers).

Recipes are only a guide. Experiment with different quantities of honey and hops - different types of honey and hops will influence the final taste.

I mentioned *Braggot* in my article on making mead. *Braggot* is a cross between mead and ale. The name is derived from the old English words brag (= malt) and gots (= honeycomb).

This recipe uses heather honey, which would give a characteristic flavour, but other honeys can be used.

Braggot

Ingredients
1 lb malt extract (not containing cod liver oil!).
1 lb heather honey.
¼ oz citric acid.
7 pints water.
Beer yeast & nutrients.

Method

Boil malt extract and honey in the water for 15 minutes, skim. Add citric acid, nutrients and, when cool, the yeast. Mature for 3 months. Best served slightly chilled.

The Ancient Laws of Wales (codified by Howell the Good 918 A.D.) stated that every free *maenol* (village) would provide the King with a vat of mead or *two vats of bragot.*

The following recipes are a little more complicated and contain more ingredients - these beers will have more body.

Honey Beer

Ingredients
½ lb malt extract.
1 lb honey.
¼ lb cracked crystal malt.
½ handful of barley or flaked rice.
½ oz hops.
¼ tsp salt.
Juice of ¼ lemon.
Beer yeast & nutrient.
Water to 1 gal.

Method

Crack crystal malt with a rolling pin. Boil crystal malt, barley or rice and hops in water for 45 minutes. Strain and wash (*sparge*) the residue with warm water. Make up the liquor to 1 gallon. Ferment etc.

Barley Mead

Ingredients

1¾ lb crushed pale malt.
2 oz flaked barley, rice, oats or maize.
1 oz hops.
1 lb honey.
1 level tsp gypsum (calcium sulphate).
1 level tsp citric acid.
½ tsp salt.
Wine yeast (Champagne, if available) & nutrient.

Method

Crush pale malt and flakes (the *grist*) and transfer to a large saucepan. Heat some water to 68 °C (152 °F) and cover the grist. Maintain this temperature for 2 hours. Strain off the solids. Return the wort to the saucepan, add the gypsum and hops - boil for 45 minutes. Strain, add remaining ingredients when cool. Skim of any scum that forms. After 3-4 days, transfer to 1 gal demijohn and fit air lock. Rack 4-6 weeks later. When fermentation has finished and the barley mead is clearing, bottle and prime. Drinkable after 4-6 weeks but will improve with keeping. **Drink from small glasses (nips) and not pints.**

Stout

Ingredients
1 lb honey.
4 oz black malt.
2 oz flaked barley.
4 oz crushed crystal malt.
8 oz malt extract.
¾ oz hops (*Fuggles* if available).
½ tsp citric acid.
½ tsp salt.
Beer yeast & nutrient.

Method

Place black malt, flaked barley, crushed crystal malt and malt extract into a saucepan with 4 pints of water. Heat to 65 °C (150 °F) and hold at this temperature for ¾ hour. Strain and add the hops to the wort. Simmer for 1½ hours. Strain over honey, citric acid, salt and nutrient. Make up to 1 gallon and add yeast when cool.

The following beers could be a talking point at your next party!

Nettle Beer

Ingredients
1 gal nettles (not pressed down).
4 oz crushed crystal malt.
1 lb malt extract.
½ lb honey.
½ oz bruised ginger.
1 tsp citric acid (or juice of 1 lemon).
¼ tsp salt.
Small handful hops.
Beer yeast.

Method

Pick young nettle tops (wear gloves!). Simmer with crystal malt for 35 minutes - add hops and boil for a further 5 minutes. Strain, squeeze nettles, make up to 1 gallon. Add remaining ingredients.

Dandelion Beer

Wash and clean 2 dandelions (roots, leaves & flowers). Use in place of nettles in above recipe.

Cheat's Honey Beer

Buy a beer kit - substitute honey for the sugar. A lager brew works well - being lightly hopped, the honey flavour comes through.

A bottled beer brewed by the now defunct brewery Steward and Patterson in Norwich was primed with honey. Paul Simpson, head brewer at Ward's Brewery in Sheffield (part of Vaux Breweries of Sunderland) remembering the taste decided to produce a honey beer (rather than just priming with honey). The result was *Waggle Dance* - a light, golden-yellow, premium beer at 5% abv. Unfortunately, the honey comes from Mexico. There are other commercial honey beers available.

Having made your mead or honey-based drink, you should, perhaps, follow the ancient law of Ireland: *There are three things in the court which must be communicated to the king before they are made known to any other person: first, every sentence of the judge; second, every new song, and third,* ***every cask of mead.***

A few more snippets ...

In the Icelandic Prose Edda the story is told of Kvasir, who wandered the world dispersing wisdom. He was murdered by two dwarfs, who mingled his blood with honey and made mead. Whoever drank the blood and honey mead became a poet and scholar. In the myth the mead is drunk by Odin, who became the patron of poets. Nordic mythology contains many stories of gods giving goddesses draughts of mead to lower their resistance to the gods' carnal intentions! One story tells of Odin, unable to make mead himself, stealing mead from Suttung. Later, Gunnlod (Suttung's daughter) gave him several draughts of mead - as he succumbed to the aphrodisiac properties of the mead, it gave him the gift of poetry and composition!

At the Norse Yuletide festival, cups of mead were drunk to honour Odin and the early Christian Church found itself unable to stop the custom. According to tradition, St Wenceslas (he who *"last looked out on the feast of Stephen"*) resolved the problem by exchanging Odin for the Archangel Michael. An early Christian prayer reads: *Let us drink this cup of mead in honour of the holy Archangel Michael beseeching him to introduce our souls to the peace of eternal life.*

In conclusion, the sad and tragic story of the Norse king Fjolne. A great feast was prepared for him by a prince called Frode, who had built a great timber mead tank. Fjolne dined well and had to be carried to his bed in one of the lofts. In the middle of the night, the story relates, he went out into the gallery to seek a certain place. On his way back he stumbled into the wrong loft and fell into the mead tank and was drowned. It illustrates the fact that you can drink mead almost indefinitely provided you remain seated. As an old Polish proverb puts it: *Mead makes you drunk from the waist down.*

Wacht heil!

If you *surf the Internet* you will find a wealth of information about mead, honey beer, and their making . There seems to be much interest in America.

Just remember that mead is reputed to be an aphrodisiac. I Have murly finshed thiss Artackel aNd aaaaa Bootie orf Me oWn MEED - CHAIRS!

Brian ('Pollio Romulus') Dennis.

a tip from the author...

Heaven forbid that you would over indulge!
But if the worst happens, try eating some honey high in fructose such as acacia honey - the alcohol is removed. How much honey is required depends more on how much you feel you can consume!

DON'T DRINK AND DRIVE.

Bibliography

In writing Mead, More Mead and Honey Drinks and Honey Beers I have consulted the numerous books and articles I have collected over many years, hence the variation in the style of recipes. Obviously I have not tested all the recipes, but I am sure they will produce interesting drinks! Similarly, I have assumed the historically information is correct. Such is the fascination of mead that I have found it difficult to reach a satisfactory conclusion – no doubt there will be a need for me to produce another publication!

- *Wassail! In Mazers of Mead* by GR Gayre. 1948: Phillimore & Co.). Reprinted *Brewing Mead - Wassail! In Mazers of Mead* by GR Gayre and C Papazion . 1986: Boulder, CO, USA - Brewers Publication.

- *The Closet of the Eminently Learned Sir Kenelme Digby, Kt., Opened* (1669). Transcript of 128 pp. Containing recipes for mead, etc. 1983: IBRA.

- *Making Mead* by Bryan Acton & Peter Duncan. 1965: *Amateur Winemaker Publications.*

- *All About Mead* by SW Andrews. 1971: Mills & Boon, Ltd.

- *Honey Wines and Beers* by Clara Furness. *Northern Bee Books:* ISBN 0-907908-39-X.

- *Mead - Making, Exhibiting & Judging* by Dr HRC Riches. 1997: *Bee Books New & Old.* ISBN 0-905652-41-X.

- *Making Cider* by Jo Deal. 1976: *Amateur Winemaking Publications.* ISBN 0 900841 45 1.

- *A Sip Through Time* by Cindy Renfrow. 1995: ISBN 0-9628598-3-4.

- *Winemaker's Companion* by BCA Turner & CJJ Berry. 1960: ISBN 263-51732-2.

- *First Steps in Winemaking* by CJJ Berry. 1982: *Amateur Winemaker Publications.* ISBNO 900841.

Lightning Source UK Ltd.
Milton Keynes UK
05 April 2011

170366UK00001B/38/P